# SCIENCE AND ENGINEERING CONFERENCE PROCEEDINGS: A GUIDE TO SOURCES FOR IDENTIFICATION AND VERIFICATION

Compiled by the
ALA/ACRL STS Task Force on Proceedings
Committee on Bibliographic Access
Edited by
Barbara DeFelice, Chair

Members

Joyce Ogburn
Colleen Power
Frederick Sepp
LeAnn Weller

Published by the Association of College and Research Libraries
A Division of the American Library Association
50 East Huron Street
Chicago, IL 60611
1-800-545-2433

ISBN 0-8389-7790-1



Printed on recycled paper.

Printed in the United States of America.

## Table of Contents


1. Preface 1

2. Contributors 2

3. Introduction 3

4. Verifying Conference Papers and Proceedings 5

5. Obtaining Conference Papers and Proceedings 9

6. References Cited 11

7. Annotated List of Sources 12

8. Title and Subject Index to List of Sources 76

9. Bibliography 79

## 1. Preface

The ALA/ACRL Science and Technology Section established a Task Force on Proceedings in 1983 to investigate some of the problems and issues involved with that important but difficult part of the scientific literature, conference papers and proceedings. The Task Force created committees to investigate different aspects of conference proceedings. The Finding Aids to Conference Proceedings Committee was charged with developing a list of tools useful to librarians needing to verify and locate conference papers and proceedings. This committee later changed its name to the Committee on Bibliographic Access. The committee developed a list of bibliographic abstracting and indexing services that cover conference proceedings, and included information on how the conference papers and proceedings are covered in these sources. That list has been incorporated into this guide as Section 7, Annotated List of Sources.

The text of this publication describes the basic problems with verifying conference proceedings and offers suggestions for handling some of these problems. It is intended for practicing librarians and library science students interested in the scientific and technical literature.

## 2. Contributors

Many people contributed information for the annotations and developed the text. Those primarily responsible for the initial drafts of each section of the text are:

Colleen Powers, for the Introduction and the section titled Verifying Conference Papers and Proceedings

Joyce Ogburn, for the section titled Obtaining Conference Proceedings and Papers

The committee has gone through several membership changes, and would like to acknowledge the contributions of the past members, and those who contributed without being formal members.

Nirmala Bangalore
William Baer
John Ferrainolo
Win Gelenter
Joyce Howard
Barbara Kautz
Ann Moore
Earl Mounts
Laura Osegueda
Diane Reiman
Martin Smith
Rebecca Uhl
Vicky Young

Special thanks to Joan Lussky who assisted in editing the text.

Thomas Mead developed alternative layouts for the annotated list, and input much of the original data.

The committee would also like to acknowledge the support of the ACRL STS in funding the work involved with inputting the data and designing the format for the annotated bibliography.

## 3. Introduction

Elusive. Ephemeral. Frustrating. Critical. Conference proceedings are perhaps better defined by these adjectives than by any AACR2R definition. The overwhelming fact is, conferences are held, papers are delivered, but only two-thirds of all conferences ever have published proceedings. Of these conferences, some or all of their papers may appear together in monographs, journals, series, or even in that elusive format the technical report (Allredge 1982).

Given the elusiveness of conference proceedings, one may seriously question whether conference papers are of great value. The American Library Association's Association of College and Research Libraries Science and Technology Section (ACRL STS) Task Force on Conference Proceedings, Committee on the Importance of Conference Proceedings was established in 1984 to examine this question. This Committee's survey of science faculty at five universities in the United States indicated a highly significant level of interest in conference papers (ALA ACRL 1986). This research clearly demonstrates that academic library users value conference papers and proceedings. Also, conference papers are frequently cited in the bibliographies of published papers, although often incorrectly and incompletely. In order to identify, verify and obtain these papers, librarians first need to understand the origin and nature of conference literature.

Exactly how do conference proceedings originate? Usually, a scientific meeting, convention or conference covers a range of topics within a subject area. Attendees participate in seminars, discussion groups, lectures, and /or poster sessions. The oral presentations at these meetings are called conference papers. The collective group of conference papers is called the conference proceedings. The organizers or sponsors of these conferences do not use consistent terminology when referring to these meetings. Other commonly used terms include symposia, colloquium, congress, discussion, institute, seminar, meeting, session, summer school, or workshop. The overall planning may be sponsored by a governmental agency, society, company, university, consortia, or even by single individuals.

Conference papers are of two types: preconference or postconference. Generally preconference materials are the abstract or written text of the planned oral presentations. Postconference materials are the edited texts written for formal publication of the conference proceedings (Bell 1982). Therefore, postconference papers are more easily identified and obtained. Preconference abstracts and other information are becoming easier to obtain because many organizations are using the Internet to post notices of current and upcoming meetings, both those sponsored by the organization and those in related subject areas. In many cases, abstracts of papers to be presented are available, along with the program, lists of presenters, registration and other information.

A study of an Optical Society of America conference reveals both the characteristics of the participants and the nature of the information produced from these conferences.

1. The attendees have significant academic credentials.

2. Research and development are the primary areas of work for the participants.

3. Most presentations deal with specific research projects, with one quarter of them consisting of a summary of current research.

4. The work presented is timely, done within one year of the meeting.

5. Three-fourths of authors plan to publish their research formally as in journal articles. (National Academy of Sciences 1969)

An important advantage to conference papers and proceedings is that they provide the first or only place where a particular research project or result is presented. The delivered papers are timely, research oriented, and state of the art. The disadvantages are that only two-thirds of conferences ever have published proceedings, only three-fourths of the people who present conference papers plan to have their work formally published, and the lengthy editorial and publishing process may negate the timeliness of the information in the conference paper.

Due to the difficulties of dealing with conference papers and proceedings, librarians need to understand the special tools and techniques that are most helpful for recognizing, identifying, locating, and acquiring these documents. The ACRL STS Task Force on Conference Proceedings, Committee on Bibliographic Access has developed this guide and annotated list of sources as a working tool for the practicing librarian involved with public services for the scientific and engineering community. It also could be used by the library school student as part of a course of study in science and technology librarianship.

## 4. Verifying Conference Papers and Proceedings

Ideally, a library user requesting a conference paper or proceedings would provide a complete, correct citation. This would include the author and title of the paper, and the name, sponsor(s), title, editor, publisher, location and date of the conference. However, this level of information is seldom provided and the librarian must attempt to identify the missing pieces and check the citation for accuracy in order to obtain the paper or proceedings for the user or the collection. This search for the conference paper information can be facilitated by answering 4 questions:

1. Is the conference paper correctly cited?
2. Was or will the conference paper or proceedings be published?
3. In what format was the conference proceedings published?
4. Is the paper available only as an abstract?

**1. Is the conference paper correctly cited?**

Many possibilities exist for errors. In order to verify that a paper has been correctly cited, consider the following possibilities:

a. The name of the author of the paper is often misspelled.

b. The conference name and/or title is incorrect, or most likely incomplete.

c. The conference date has been misstated, or the volume or pagination is incorrect.

A common problem is a misspelled author's name. Therefore, this is a good place to begin verifying a citation. Any cursory check of an index will show the extent of the mistyped author's name. Author searches in the author indexes of printed subject-specific indexes such as *Chemical Abstracts* or a general science index such as *Science Citation Index* will probably prove the least expensive and the least time consuming method of verifying the author. This process also provides additional information for establishing the complete citation.

The accepted name of a conference and the title of its proceedings can be a challenge to determine even when you have the published proceedings right in front of you. With multiple titles, cosponsors, and series names on the cover, the title page, and the spine, the writer may have picked any one or variation of these to use in the citation. Just as with the spelling of an author's name, there are many chances for an error in the pagination, volume and year. Indexes to conference proceedings and conference papers can be helpful in verifying the correct citation.

The principal indexes in most disciplines cover conference proceedings and papers, and can be searched at the paper level. However, there are some limitations in conference coverage in many indexes. For instance, ISI's *Science Citation Index* excludes all proceedings which give only the titles or abstract of papers. *Biological Abstracts/Reports, Reviews and Meetings* only began listing those types of proceedings in the 1980's.

The major general indexes that offer conference paper indexing as well as access to conference proceedings information that should be consulted include *Conference Papers Index* and *Index to Scientific and Technical Proceedings. Proceedings in Print* and *Directory of Published Proceedings* cover complete conferences, and are useful in determining if and where a conference proceedings was published. The annotated bibliography of indexes and abstracts in Section 7 is designed to allow the librarian to easily select the most useful tool for identification and verification of conference paper citations, depending on the subject area.

For current conferences, the sponsoring organization may have an Internet site which will list presenters, paper titles and abstracts. A good example of this kind of information source is the American Mathematical Society's E-Math Home Page. This World Wide Web site's address is: http://e-math.ams.org/. Other examples of Internet sites which provide current or upcoming conference paper abstracts are the American Astronomical Society's WWW site: http://blackhole.aas.org, and the Association for Computing Machinery Design Automation Special Interest Group's site: http://kona.ee.pitt.edu/ . The Lunar and Planetary Science Institute's Conference Program and Abstracts can be searched by connecting via Telnet to: LPI.JSC.NASA.GOV. Sources of information like these for other subject areas or organizations can be located by using Gopher searchers to locate an organization's name or browsing the World Wide Web by topic or organization.

The major bibliographic utilities, OCLC, RLIN and WLN, are an excellent source for verification of the name, title, editor, and place of published conference proceedings. The documentation for searching the utilities should address the special techniques that work best for conference proceedings.

**2. Was the conference proceedings published or will it be?**

If the conference paper or proceedings cannot be found in the indexes, it is possible that it was not published. Verifying this can be a time-consuming process. Several methods are listed below.

The sponsoring organizations' publications catalogs may aid in determining if the conference was published. The *Encyclopedia of Associations* includes information about conferences sponsored by an association, and gives the association's phone number. Some times several calls to different offices, editors and former editors may be placed before the necessary information is obtained because many associations have poor bibliographic control or access to the publications of their various divisions, study groups or councils. An organization's Internet Gopher or WWW site may include publications available or information about where its conference papers were published.

Contacting the paper's author is another way to determine if a conference was published. Many print and online indexes, such as *Science Citation Index*, include the author's address. Other sources for address information would include: *The National Faculty Directory* and *American Men and Women of Science*. Should these directories fail, check the major associations membership directories such as the *American Mathematical Society Membership Directory* or the *IEEE Membership Directory*. Some of these directories, such as the *American Mathematical Society's Combined Membership List*, are now available on the Internet.

**3. In what format was the conference proceedings published?**

The principal formats for a published conference proceedings include: numbered series, unnumbered irregular series, monographs in series, special numbers or issues or supplements of journals, and separate monographs available from standard trade publishers. The librarian may find the bibliographic utilities helpful in identifying the bibliographic format(s) of a conference proceedings.

**4. Is the paper available only as an abstract?**

While no figures have been found to identify the percentage of papers available in abstract form only, clearly this is a relatively common occurrence. Fortunately, most printed and online sources will indicate when a paper is available as an abstract only. One way of obtaining the desired information is to find something else by the author that expands on the original topic and was published close to the same time as the conference was held. Many papers eventually are reworked, modified and edited to appear later in other scholarly formats. Another method is to locate the author by using similar sources to those described in the preceding paragraphs, and to request a copy of the paper directly from the author. This should satisfy the patron's needs for currency of research.

## 5. Obtaining Conference Papers and Proceedings

Just as proceedings can be difficult to locate or verify by the reference librarian, proceedings can cause problems for the acquisition and interlibrary loan departments of the library. Problems include verification, simultaneous publication as a serial and a monograph, simultaneous publication by two or more publishers, lack of mention of the conference and publication under different titles. For example, IEEE and ACM have been known to publish conference proceedings jointly, in addition to the conference being published by a commercial publisher.

Depending on library policy, several options exist for acquiring a conference paper or the entire proceedings: 1) purchasing a copy for the patron; 2) ordering a copy of the paper; 3) ordering a copy of the entire proceedings; 4) soliciting a conference proceedings from an attendee as a gift. The following paragraphs will outline the different methods for obtaining conference proceedings.

If the citation to a conference paper has been found through an online search service with document delivery capability, the easiest way to obtain a copy is to place an online order directly through the vendor. The cost is charged to the search account and delivery time is minimal. Written documentation for databases includes information regarding document delivery procedures. This service offers the library the option of giving the conference paper directly to the patron upon receipt and passing the charges along to the patron.

The library may wish to order the proceedings through one of several services which handle a great many conference publications. The most common are Interdok, NTIS, Information on Demand, and the U.S. Government Printing Office. Of course some government publications may be available through the federal library deposit program. NTIS allows a library to establish a profile of subjects and agencies for publications which are desired. NTIS, Information on Demand, and Interdok offer online ordering services in additional to traditional ordering methods. These services allow the library to search their database for the correct title and then to place an order.

Many vendors of library materials can readily obtain conference proceedings, regardless of whether they are published by a commercial publisher, a society, or a government agency. Often these materials are obtained at a higher cost to the library, for no discounts are given to the vendor by the publisher. The advantage of ordering from a vendor may include consolidating orders, easy claiming, no prepaying. and dealing with a trusted supplier.

Ordering direct from the publisher may seem the simplest way to obtain conference proceedings, though this may cause some problems for the library. This may require a prepayment, a special purchase order, more follow up, and other special efforts and correspondence by the library. The advantages are that the publisher can frequently give the quickest turn around time on the order and may not add extra charges to send the material.

Many libraries have programs through which they receive all or carefully selected publications, which may include conference proceedings. Often a library will establish a membership with an organization, allowing the purchase of publications at a discount and perhaps guaranteeing their automatic receipt. Since associations and societies are major publishers of conference proceedings, a membership can gain the library easier access to these publications. In addition, many major scientific publishers produce conference proceedings and standing orders may be placed for these materials. If desired, a library may be able to include conference proceedings within its approval plan profiles.

In sum, conference proceedings can be acquired using traditional acquisitions methods. They may demand some special verification procedures in order to avoid duplicate purchase when a conference is published in several different ways.

## 6. References Cited


Allredge, Shirley. "Techniques and strategies used by seasoned librarians for the identification and verification of incorrect or incomplete conference citations." In Workshop on Conference Literature: Conference Literature: Its Role in the Distribution of Information, Proceedings , 110-117, 1982.

Bell, Helen C. "Acquiring conference papers through interlibrary loan service."In Workshop on Conference Literature: Conference Literature: Its Role in the Distribution of Information, Proceedings , 118-147, 1982.

National Academy of Sciences. Committee on Scientific and Technical Communication.
Scientific and Technical Communication: A pressing national problem and recommendations, pp. 107-110, 1969.

American Library Association. Association of College and Research Libraries. Science and Technology Section. Taskforce on the Importance of Conference Proceedings. Report on the Pilot Study on the Value of Conference Proceedings in the Scientific Literature, pp. 1-58, 1986.

## 7. Annotated List of Sources

Indexes to periodicals are a major source of bibliographic information. However, the few indexes devoted to conference papers are not exhaustive, so subject specific indexes and abstracts must also be consulted. For these to be used most effectively, it is necessary to understand what conference literature the index covers and how it is indexed.

This annotated list includes the major abstracting and indexing services in science and engineering that were being published in 1994 and cover conference proceedings. Medical indexes are excluded. The indexes and abstracts in this list were selected because they are useful for verifying and identifying science and engineering conference papers and proceedings. This is not a comprehensive list.

Both electronic and print versions of many sources are listed. It is difficult to provide exact information about how each database and vendor handles conference papers and proceedings in various electronic formats such as online or CD-ROM, as different vendors of a database may organize and index the materials in very different ways. Some institutions make selected indexes available through their local OPAC and searchable by the same interface. How conference proceedings are handled in these cases will depend on local practice.

Since publishers, titles and form of publication change frequently, not all the information here may be up-to-date. The list was last corrected and updated in 1994. More and more titles are available on CD-ROM, as well as online. It is possible that organizations which now list author, title and abstract information about current conference papers through the Internet will expand to list complete papers.

Publishers of the indexing services were contacted to ascertain their policies governing the inclusion and indexing of conference literature. Information from these policies is included where available.

The list is followed by a title and subject index. It is hoped that this list will give practitioners a place to start when verifying and locating conference papers and proceedings.

## Agricultural Engineering Abstracts

| | |
|---|---|
| **Producer** | C.A.B. International<br>(Commonwealth Agricultural Bureaux)<br>845 North Park Ave.<br>Tucson, AZ 85719 |
| **Subjects** | Agriculture |
| **Publication Dates** | 1976 |
| **Access Points** | Author, Subject, Broad subject divisions in 'Contents' pages |
| **Conference Indexing** | Conferences are listed in 'Contents' and individual papers are listed in the annual subject index under the heading 'Conferences' or ' Conference Proceedings', as well as listed by subject, and author in the appropriate indexes. |
| **Conference Coverage** | States in beginning of some issues that they select "significant items". Papers from conferences appear to be covered in highly selective fashion, usually only one or two papers from a conference. Several different CAB journals may cover the same Conference, selecting different papers from each conference for inclusion in the various journals. |
| **Abstracts Note** | All materials that are included are abstracted, including conferences and conference papers |
| **Ordering Information** | Many issues include a form and detailed instructions for ordering photocopies of listed papers from C.A.B.'s Document Delivery Service |
| **Electronic Equivalent** | Available online as CAB Abstracts, accessible as a subfile in the database from 1972- on BRS, Dialog, STN. Available on CD-ROM 1984- through SilverPlatter. |
| **Notes** | The journals are prepared centrally, but indexed by each separate CAB Institute. |

## Animal Breeding Abstracts

| | |
|---|---|
| **Producer** | C.A.B. International (Commonwealth Agricultural Bureaux)<br>845 North Park Ave.<br>Tucson, AZ 85719 |
| **Subjects** | Agriculture |
| **Publication Dates** | 1933- |
| **Access Points** | Author, Subject, Broad subject divisions in 'Contents' pages |
| **Conference Indexing** | Conferences are listed in 'Contents' and individual papers are listed in the subject index under the heading 'Conferences' or 'Conference Proceedings', as well as listed by subject, and author in the appropriate indexes. |
| **Conference Coverage** | States in beginning of some issues that they "select significant items". Papers from conferences appear to be covered in highly selective fashion, usually only one or two papers from a conference. Several different CAB journals may cover the same Conference, selecting different papers from each conference for inclusion in the various journals. |
| **Abstracts Note** | All materials that are included are abstracted, including conferences and papers. |
| **Ordering Information** | Many issues include a form and detailed instructions for ordering photocopies of listed papers from C.A.B.'s Document Delivery Service. |
| **Electronic Equivalent** | Available online as CAB Abstracts, accessible as a subfile in the database from 1972- on BRS, Dialog, STN. Available on CD-ROM 1984- through SilverPlatter. |
| **Notes** | The journals are prepared centrally, but indexed by each separate CAB Institute. |

## Apicultural Abstracts

| | |
|---|---|
| **Producer** | C.A.B. International (Commonwealth Agricultural Bureaux)<br>845 North Park Ave.<br>Tucson, AZ 85719 |
| **Subjects** | Agriculture, Bees, Entomology |
| **Publication Dates** | 1950- |
| **Access Points** | Author, Subject, Broad subject divisions in 'Contents' pages |
| **Conference Indexing** | Conferences are listed in 'Contents' and individual papers are listed in the subject index under the heading 'Conferences' or 'Conference Proceedings', as well as listed by subject, and author in the appropriate indexes. |
| **Conference Coverage** | States in beginning of some issues that they "select significant items". Papers from conferences appear to be covered in highly selective fashion, usually only one or two papers from a conference. Several different CAB journals may cover the same Conference, selecting different papers from each conference for inclusion in the various journals. |
| **Abstracts Note** | All materials that are included are abstracted, including conferences and conference papers |
| **Ordering Information** | Many issues include a form and detailed instructions for ordering photocopies of listed papers from C.A.B.'s Document Delivery Service |
| **Electronic Equivalent** | Available online as CAB Abstracts, accessible as a subfile in the database from 1972- on BRS, Dialog (file 50), STN. Available on CD-ROM 1984- through SilverPlatter. |
| **Notes** | The journals are prepared centrally, but indexed by each separate CAB Institute. |

## Applied Science and Technology Index

| | |
|---|---|
| **Producer** | H. W. Wilson Company<br>950 University Avenue<br>Bronx, New York 10452 |
| **Subjects** | Engineering, Materials Science |
| **Publication Dates** | 1957 - |
| **Access Points** | Subject index. Monthly publication with quarterly and annual cumulations. |
| **Conference Indexing** | The conference as a whole is indexed by subject. The name of the sponsoring organization may also be an indexing term. Individual papers may or may not also be indexed by additional subject terms. Access online is also by title word "proceedings", and article contents code "sympo". |
| **Conference Coverage** | All conference proceedings included in the approximately 335 journals covered by this index. |
| **Abstracts Note** | No abstracts are provided by this Wilson index. However if abstracts of conference papers are included in a journal, "abstracts of papers" is added to title. |
| **Ordering Information** | H. W. Wilson has no document delivery service. |
| **Electronic Equivalent** | Available online since 1983 through WILSONLINE and OCLC EPIC. Available on CD-ROM 1983- through WilsonDisk and SilverPlatter. |
| **Notes** | Approximately 484 conference proceedings have been added to the online database as of August 1, 1990. |

## Aquatic Sciences & Fisheries Abstracts

| | |
|---|---|
| **Producer** | Cambridge Scientific Abstracts<br>7200 Wisconsin Avenue<br>Bethesda, MD 20814 |
| **Subjects** | Biology, Agriculture, Fish and Fisheries, Environmental Studies |
| **Publication Dates** | 1971- |
| **Access Points** | Author, Subject, Taxonomic, Geographic |
| **Conference Indexing** | Individual papers are cited in this index instead of using one record for the whole conference. In the online database, document type is a searchable field. |
| **Conference Coverage** | Coverage of conferences is about 22% of citations in this index. English is the predominant language but coverage is international. |
| **Abstracts Note** | Almost all citations have abstracts. |
| **Ordering Information** | Cambridge Scientific Abstracts/ASFA suggests the British Library Lending Division, Information on Demand (IOD), the National Agricultural Library, and the National Technical Information Service. |
| **Electronic Equivalent** | Available online 1978- on STN as AQUASCI and Dialog (file 44). Available on CD-ROM 1978- from SilverPlatter and Cambridge Scientific Abstracts. |
| **Notes** | Part 1=Biological Sciences and Living Resources<br>Part 2=Ocean Technology, Policy and Non-Living Resources<br>Part 3=Aquatic Pollution and Environmental Quality (1990-) |

## Artificial Intelligence Abstracts

| | |
|---|---|
| **Producer** | Bowker A & I Publishing<br>R.R.Bowker<br>245 West 17th St.<br>New York, New York 10011 |
| **Subjects** | Computer Science |
| **Publication Dates** | 1985 - |
| **Access Points** | Author, Subject, Industry, Source |
| **Conference Indexing** | Document types such as 'conference proceedings' or 'conference paper' are indicated on each record, but not indexed as such. A separate section "Conferences and Events" provides dates and locations of selected conferences. In the online databases, document type is a searchable field. |
| **Conference Coverage** | Core conference sources are scanned. Conference papers make up 30% in the energy and environment indexes to 50% in the computer indexes of the items in these Bowker indexes. Only English language materials are included. |
| **Abstracts Note** | Abstracts are included for all citations. |
| **Ordering Information** | Document Retrieval Service makes documents available in hard copy or microfiche. |
| **Electronic Equivalent** | Was the Supertech file on Dialog; no longer available. |
| **Notes** | Supertech includes records from two discontinued titles, Telecommunications Abstracts and Telegen Abstracts. Supertech in Dialog includes Artificial Intelligence Abstracts, CAD/CAM Abstracts, Robotics Abstracts. |

## Astronomy and Astrophysics Abstracts

| | |
|---|---|
| **Producer** | Astronomisches<br>-Institut Heidelberg<br>W-6900 Heidelberg 1<br>Germany |
| **Subjects** | Astronomy, Astrophysics |
| **Publication Dates** | 1969- |
| **Access Points** | Author, Subject, Object |
| **Conference Indexing** | A separate section is called "Proceedings of colloquia, congresses, meetings and symposia". This lists the monographic entries for the proceedings under which is given the entry numbers for the abstract and citation of each separate paper. The citation for each paper in turn refers to the citation for the whole conference. All conferences are not indexed completely; the papers must be within the subject scope of Astronomy and Astrophysics Abstracts. Document type 'conference' is a searchable field in the online database. |
| **Conference Coverage** | Worldwide coverage. |
| **Abstracts Note** | All individual conference papers have abstracts, but the whole conference citations do not. |
| **Ordering Information** | |
| **Electronic Equivalent** | Was available online 1979-1994 as the PHYS file on STN. Will be included in the INSPEC file 1995- through Dialog and STN. |
| **Notes** | |

## Bibliography and Index of Geology

| | |
|---|---|
| **Producer** | American Geological institute<br>4220 King St.<br>Alexandria, VA 22302 |
| **Subjects** | Geosciences |
| **Publication Dates** | 1969-, although continues other titles covering the geosciences from 1785 |
| **Access Points** | Author, Subject, Broad fields of interest |
| **Conference Indexing** | The title of each conference proceedings indexed is listed separately in the front of the print version. The whole conference proceedings and each separate paper are indexed. Full proceedings are listed under Symposia in the front of the printed index, and all authors and editors, the proceedings title, location and date are included in the record for each paper. In the online version, the record for each conference is indicated as a monographic record and the record for each paper is indicated as an analytic record. Records for both the whole conference and each paper are designated "conference publication" as the document type in the online version. |
| **Conference Coverage** | Any full conference paper relating to geology is indexed regardless of language. Abstracts of conference papers are indexed if it is known that the full papers will not all be published in a proceedings volume. An example is the *Geological Society of America Abstracts with Programs*. Each abstract from this title is indexed. Generally, only English-language abstracts of conference papers are indexed. All conference proceedings acquired by the U.S. Geological Survey library in Reston, Virginia are indexed. Some organizers of the meetings listed in the Calendar of Meetings in *Geotimes* are contacted with a request for conference proceedings. |

| | |
|---|---|
| **Abstracts Note** | No abstracts of the citations |
| **Ordering Information** | AGI provides document delivery services. |
| **Electronic Equivalent** | Available online 1785- GEOREF file on Dialog (file 89), ORBIT and STN. Available on CD-ROM 1785- through SilverPlatter. |
| **Notes** | Conference proceedings follow journal articles in priority of indexing regardless of language or country of origin unless the conference paper was published as an article in a core journal, in which case it gets priority indexing. |

## Bibliography of Agriculture

| | |
|---|---|
| **Producer** | National Agricultural Library<br>10301 Baltimore Blvd.<br>Beltsville, MD 20705<br><br>Oryx Press ( publisher)<br>4041 N. Central Indian School Rd.<br>Phoenix, AZ 85012-3397 |
| **Subjects** | Agriculture |
| **Publication Dates** | 1942- |
| **Access Points** | Personal and Corporate Author, Subject, Monthly and annual cumulations. |
| **Conference Indexing** | Monographic conference proceedings are listed under the subject heading "conferences". There is one record for the whole conference and separate records for all papers within the proceedings. |
| **Conference Coverage** | All conference proceedings acquired and cataloged by the National Agricultural Library are indexed. Cover to cover indexing is provided for every conference. |
| **Abstracts Note** | No abstracts for conferences. |
| **Ordering Information** | Interlibrary loan from the National Agricultural Library |
| **Electronic Equivalent** | Available online 1970- as AGRICOLA on BRS and Dialog (file 10). Available on CD-ROM 1970- through SilverPlatter. |
| **Notes** | Subject index terms are from the *C.A.B. Thesaurus* |

## Biological Abstracts/RRM
## (Reports, Reviews and Meetings)

| | |
|---|---|
| **Producer** | BIOSIS<br>2100 Arch Street<br>Philadelphia, PA 19103-1399 |
| **Subjects** | Biology, Biochemistry |
| **Publication Dates** | 1965- |
| **Access Points** | Author, Title, Subject, Biosystematic, Generic |
| **Conference Indexing** | BIOSIS uses various terms for these publications : meeting, conference, symposium, proceedings, colloquium and congress. BIOSIS provides one record for the whole conference and separate records for specific papers. Conference citations appear in the "Meetings" section of RA/RRM. In the online database, the Concept Code 00520 is assigned to each record. |
| **Conference Coverage** | All areas of the life sciences are considered. Material is selected on the basis of subject, geographic scope, timeliness, size of the meeting, previous BIOSIS coverage, format (abstracts or papers) and special requests. About one-third of the items added in 1988 was conference literature. There is no stated language restriction. |
| **Abstracts Note** | Abstracts for all citations. |
| **Ordering Information** | BIOS IS suggests the John Crerar Library, Medical Library Center of New York, National Agricultural Library, Linda Hall Library, The British Library (Lending Division), The Royal Netherlands Academy of Arts and Sciences Library, and University Microfilms International. |
| **Electronic Equivalent** | Available online 1969- as BIOSIS Previews (includes Biological Abstracts and BA/Reports, Reviews and Meetings) on BRS, Dialog (file 5) and STN. Available on CD-ROM 1989- through SilverPlatter. |
| **Notes** | Previous title: BioResearch Index (1965-1979) |

## Biological and Agricultural Index

| | |
|---|---|
| **Producer** | H. W. Wilson Company<br>950 University Avenue<br>Bronx, New York 10452 |
| **Subjects** | Biology Agriculture |
| **Publication Dates** | 1964-present. From 1916-1963 named Agriculture Index. Monthly, with quarterly and annual cumulations. |
| **Access Points** | Subject index primarily. Topical subject terms and names of societies, associations, etc., are used as headings. |
| **Conference Indexing** | Conference proceedings are indexed by subject and by name of society or association. Individual papers are also indexed. Each conference proceeding is assigned an article contents code "sympo" for easy online retrieval. |
| **Conference Coverage** | All conference proceedings appearing in the 226 journals indexed by BAI are covered thoroughly. |
| **Abstracts Note** | Symposium proceedings that contain abstracts of papers are indexed and a title enhancement "abstracts of papers" is added. |
| **Ordering Information** | H. W. Wilson has no document delivery service. |
| **Electronic Equivalent** | Available 1983- on WILSONLINE and OCLC EPIC. Available on CD-ROM 1983- through WilsonDisk and SilverPlatter. |
| **Notes** | About 1300 conference proceedings (workshops, symposia, etc.) have been indexed as of August 1, 1990. |

## CAD/CAM Abstracts

| | |
|---|---|
| **Producer** | Bowker A & I Publishing<br>R.R.Bowker<br>245 West 17th St.<br>New York, New York 10011 |
| **Subjects** | Computer ScienceEngineering |
| **Publication Dates** | 1985 - |
| **Access Points** | Author, Subject, Industry, Source |
| **Conference Indexing** | Document types such as 'conference proceedings' or 'conference paper' are indicated on each record, but not indexed as such. A separate section 'Conferences and Events' provides dates and locations of selected conferences. In the online databases, document type is a searchable field. |
| **Conference Coverage** | Core conference sources are scanned. Conference papers make up 50% of the items in this index. Only English language materials are included. |
| **Abstracts Note** | Abstracts for all citations. |
| **Ordering Information** | Document Retrieval Service makes documents available in hard copy or microfiche. |
| **Electronic Equivalent** | Was 1985-in the Supertech file on Dialog; no longer available. |
| **Notes** | Supertech includes records from two discontinued titles, Telecommunications Abstracts and Telegen Abstracts. Supertech in Dialog includes Artificial Intelligence Abstracts, CAD/CAM Abstracts, Robotics Abstracts |

## Chemical Abstracts

**Producer**

Chemical Abstracts Service
2540 Olentangy River Rd.
P.O. Box 3012
Columbus, OH 43210

**Subjects**

Chemistry, Chemical Engineering, Biochemistry
Materials Science

**Publication Dates**

1907-

**Access Points**

Issue indexes: Keyword, Author, Patent, Volume and Collective Indexes: Author, General Subject, Chemical Substance, Formula, Patent, Numerical Patent

**Conference Indexing**

Indexing is provided for the conference as a whole, as well as the individual papers within the conference. Conferences may be given a document type: C (Conference), J (Journal), or B (Book). Editors are included for book announcements and authors are included for their papers.

**Conference Coverage**

Conference proceedings must be published in what would be considered an archival form. There must be clear evidence that a work has undergone editorial review. Proceedings must consist of complete papers and must report the results of new work. Conferences must attract participants from a large geographical area (preferably international). Meetings of local or regional scope are NOT covered. Any indication that the papers may be published elsewhere will eliminate such a proceedings from coverage.

**Abstracts Note**

A record is provided for the conference as a whole with a book announcement. Some conferences have the chemical papers abstracted. Both are indexed.

**Ordering Information**

CAS provides document delivery services and online ordering of documents. STN and Dialog offer document delivery.

**Electronic Equivalent**

Available online 1967- as CA SEARCH on Dialog (file 399), STN; 1977- on BRS, CISTI. 12th Collective Index, 1987-1991, is available on CD-ROM through CAS.

**Notes**

Conference publications are given no set priority, and the language or sponsor does not influence priority. Chemical Abstracts takes great care to see that materials are evaluated prior to indexing.

## Computer and Control Abstracts

| | |
|---|---|
| **Producer** | INSPEC<br>445 Hoes Lane<br>Piscataway, NJ 08855-1331<br>(Published by the Institution of Electrical Engineers (IEE) in association with the Institute of Electrical and Electronics Engineers (IEEE). |
| **Subjects** | Computer Science |
| **Publication Dates** | 1966- |
| **Access Points** | Personal and Corporate Author, Subject (by classification and by word), Conference Index, Book Index, Bibliography Index. Monthly and semiannual cumulations. Bibliography Index is a list of articles containing, generally, 50 or more references. Book Index is a title listing of books abstracted; abstracts of book as a whole are so indicated, for others each chapter is abstracted. Most are abstracted as a whole. Title, place, publisher, date and abstract |
| **Conference Indexing** | For most entries, one record for the conference as a whole and separate records for the papers on subjects covered by this index are provided. That is, IEEE conferences fully covered by INSPEC may not be fully covered in any one subtitle. The Conference Index is arranged by name of the conference; abstract numbers for the whole conference and for individual papers are listed. The titles of the papers appear at the abstract entry, not in this index. |
| **Conference Coverage** | Conferences published separately and those reported in covered journals are included. |
| **Abstracts Note** | All citations include abstracts. |
| **Ordering Information** | Publisher does not provide documents. They are generally available from most commercial document delivery services. |
| **Electronic Equivalent** | Available online 1969- as INSPEC on BRS, Dialog (file 2), Orbit, STN, CISTI, CAN/OLE, Data-Star, CEDOCAR, ESA-IRS, FIZ Technik, Tech Data. Available on CD-ROM 1989- through UMIOndisk as part of INSPEC. |
| **Notes** | Controlled subject vocabulary provided by the INSPEC Thesaurus. |

## Conference Papers Index

| | |
|---|---|
| **Producer** | Cambridge Scientific Abstracts<br>7200 Wisconsin Avenue<br>Bethesda,MD 20814 |
| **Subjects** | Multidisciplinary |
| **Publication Dates** | 1973- |
| **Access Points** | Subject and Author indexes to papers; Conference Locator indexes to date, location and subject category of the whole conference; Contents index by broad subject area. |
| **Conference Indexing** | Each paper within a proceeding is indexed. |
| **Conference Coverage** | All papers presented at meetings covered by Conference Papers Index are included; papers do not need to be published. The information included in Conference Papers Index is taken from selected final programs or abstracts of papers presented at conferences, supplemented by responses to questionnaires sent by the Conference Papers Index staff. |
| **Abstracts Note** | No abstracts of the citations. |
| **Ordering Information** | Includes ordering information for any abstract publications, preprints, reprint volumes, conference proceedings, or other publications that are expected to issue from the conference. The address of the primary author is included whenever available so that reprints can be requested directly. |
| **Electronic Equivalent** | Available online 1973- as Conference Papers Index on Dialog (file 77) and CONFSCI on STN. |
| **Notes** | Formerly titled "Current Programs." |

## Dairy Science Abstracts

| | |
|---|---|
| **Producer** | C.A.B. International (Commonwealth Agricultural Bureaux)<br>845 North Park Ave.<br>Tucson, AZ 85719 |
| **Subjects** | Agriculture |
| **Publication Dates** | 1938- |
| **Access Points** | Conferences listed in 'Contents.'<br>Author, Subject, broad divisional by subjects in 'Contents' |
| **Conference Indexing** | Conferences are listed in 'Contents' and individual papers are listed in the annual subject index under the heading 'Conferences' or 'Conference Proceedings,' as well as listed by subject, and author in the appropriate indexes. |
| **Conference Coverage** | States in beginning of some issues that they "select significant items". Papers from conferences appear to be covered in highly selective fashion, usually only one or two papers from a conference. Several different CAB journals may cover the same Conference, selecting different papers from each conference for inclusion in the various journals. |
| **Abstracts Note** | All materials that are included are abstracted, including conferences and papers. |
| **Ordering Information** | Many issues include a form for ordering photocopies of listed papers from CAB's Document Delivery Service, and detailed information. |
| **Electronic Equivalent** | Available online as CAB Abstracts, accessible as a subfile in the database from 1972- on BRS, Dialog, STN.<br>Available on CD-ROM 1984- through SilverPlatter. |
| **Notes** | The journals are prepared centrally, but indexed by each separate CAB Institute. |

## Directory of Published Proceedings. Series PCE: Pollution Control and Ecology

| | |
|---|---|
| **Producer** | InterDok Corp.<br>173 Halstead Ave.<br>Box 326<br>Harrison, NY 10528 |
| **Subjects** | Pollution, Ecology |
| **Publication Dates** | 1974- |
| **Access Points** | Editor of the conference proceedings, location of the conference, general subject of the conference, sponsor of the conference. |
| **Conference Indexing** | All indexing is at the full conference proceedings level. There is no indexing for individual papers. Conferences are listed chronologically by date. |
| **Conference Coverage** | All the citations are to conferences, both preprints and papers from . International coverage. Includes preprints and published proceedings. |
| **Abstracts Note** | |
| **Ordering Information** | |
| **Electronic Equivalent** | |
| **Notes** | Citations in PCE are taken from the SEMT and SSH parts of the Directory of Published Proceedings |

## Directory of Published Proceedings. Series SEMT-Science, Engineering, Medicine and Technology

| | |
|---|---|
| **Producer** | InterDok Corp.<br>173 Halstead Ave.<br>Box 326<br>Harrison, NY 10528 |
| **Subjects** | Multidisciplinary |
| **Publication Dates** | 1965- |
| **Access Points** | Editor of the conference proceedings, location of the conference, general subject of the conference, sponsor of the conference. |
| **Conference Indexing** | All indexing is at the full conference proceedings level. There is no indexing for individual papers. Conferences are listed chronologically by date. |
| **Conference Coverage** | All the citations are to conferences. International coverage. Includes preprints and published proceedings. |
| **Abstracts Note** | |
| **Ordering Information** | |
| **Electronic Equivalent** | |
| **Notes** | |

## Ecological Abstracts

| | |
|---|---|
| **Producer** | GeoAbstracts<br>Regency House<br>34 Duke St.<br>NR3 3AP<br>England |
| **Subjects** | Environmental Studies, Ecology |
| **Publication Dates** | 1974- |
| **Access Points** | Author and general subject |
| **Conference Indexing** | Most conference citations are to individual conference papers and/or whole proceedings which were published as journal articles. |
| **Conference Coverage** | About 10% of the citations are to conference papers or proceedings. A few of the *Journals Abstracted in Ecological Abstracts* are society proceedings. |
| **Abstracts Note** | All citations have abstracts |
| **Ordering Information** | |
| **Electronic Equivalent** | Available 1980-as GEOBASE on Dialog and ORBIT.<br>Available on CD-ROM 1980- through SilverPlatter. |
| **Notes** | |

## Energy Information Abstracts

| | |
|---|---|
| **Producer** | Bowker A & I Publishing<br>R.R.Bowker<br>121 Chanlon Rd.<br>New Providence, NJ 07974 |
| **Subjects** | Energy |
| **Publication Dates** | 1971- |
| **Access Points** | Author, Subject, Industry, Source |
| **Conference Indexing** | Document types such as 'conference proceedings' or 'conference paper' are indicated on each record, but not indexed as such. A separate section 'Conferences and Events' provides dates and locations of selected conferences. In the online databases, document type is a searchable field. |
| **Conference Coverage** | Core conference sources are scanned. Conference papers make up 30% in the energy and environment indexes to 50% in the computer indexes of the items in these indexes. Only English language materials are included. |
| **Abstracts Note** | Abstracts are included for all citations. |
| **Ordering Information** | Document Retrieval Service makes documents available in hard copy or microfiche. |
| **Electronic Equivalent** | Available online 1976- as Energyline on Dialog and ORBIT. Available on CD-ROM through Bowker-Reed Reference Electronic Publishing as part of Enviro/Energyline Abstracts Plus. |
| **Notes** | |

## Engineering Index

| | |
|---|---|
| **Producer** | Engineering Information, Inc.<br>Castle Rock on the Hudson<br>Hoboken, NJ 07030 |
| **Subjects** | Engineering |
| **Publication Dates** | 1884- |
| **Access Points** | Author, Subject |
| **Conference Indexing** | In the online file, 'conference paper', 'conference article' and 'conference proceedings' are all searchable as document type. There is a record for the whole conference and each individual paper. |
| **Conference Coverage** | Significant published proceedings were formerly indexed in Ei Engineering Meetings. These records (over 480,000) are included in the online file. About 2,000 conferences are covered each year. |
| **Abstracts Note** | Abstracts online for all citations, whether to an individual paper or whole proceedings. |
| **Ordering Information** | Engineering Information, Inc. provides document delivery of photocopies of articles and conference papers. |
| **Electronic Equivalent** | Available online 1970 -as Compendex on Dialog (file 8), ORBIT and STN. Available, along with Engineering Meetings, on CD-ROM 1988- through Dialog OnDisc as EI Compendex Plus. |
| **Notes** | |

## Entomology Abstracts

| | |
|---|---|
| **Producer** | Cambridge Scientific Abstracts<br>7200 Wisconsin Avenue<br>Bethesda, MD 2 0814 |
| **Subjects** | Biology, Entomology |
| **Publication Dates** | 1969- |
| **Access Points** | Author, Subject, Taxonomic |
| **Conference Indexing** | Individual papers are cited instead of using one record for the whole conference. |
| **Conference Coverage** | Conference literature is routinely identified and included; however, CSA also publishes Conference Papers Index (CPI) which is devoted to this literature. English is the predominant language but coverage is international. |
| **Abstracts Note** | 90% of all citations have abstracts since 1979. |
| **Ordering Information** | None except the author's address. |
| **Electronic Equivalent** | Available online 1978- as Life Sciences Collection (file 76) on Dialog. |
| **Notes** | CSA publishes about 30 scientific abstracts/indexes. |

## Environment Abstracts

| | |
|---|---|
| **Producer** | Congressional Information Service, Inc.<br>4520 East-West Highway, Suite 800<br>Bethesda, MD 20814-3389 |
| **Subjects** | Environmental Studies |
| **Publication Dates** | 1971- |
| **Access Points** | Author, Subject, Industry, Source, Geographic |
| **Conference Indexing** | Document types such as 'conference proceedings' or 'conference paper' are indicated on each record, but not indexed as such. A separate section "Conferences and Events" provides dates and locations of selected conferences. In the online databases, document type is a searchable field. |
| **Conference Coverage** | Core conference sources are scanned. Conference papers make up 30% in the energy and environment indexes to 50% in the computer indexes of the items in these indexes. Only English language materials are included. |
| **Abstracts Note** | Abstracts are included for all citations. |
| **Ordering Information** | Document Retrieval Service makes documents available in hard copy or microfiche. Envirofiche, a set of documents indexed can be purchased separately. |
| **Electronic Equivalent** | Available online 1971- as Enviroline on Dialog (file 40). |
| **Notes** | Absorbed Acid Rain Abstracts in 1991. |

## Field Crop Abstracts

| | |
|---|---|
| **Producer** | C.A.B. International (Commonwealth Agricultural Bureaux)<br>845 North Park Ave.<br>Tucson, AZ, 85719 |
| **Subjects** | Agriculture |
| **Publication Dates** | 1937- |
| **Access Points** | Conferences listed in 'Contents.'<br>Author, Subject, broad divisional by subjects on 'Contents' page. |
| **Conference Indexing** | Conferences are listed in 'Contents' and individual papers are listed in the annual subject index under the heading 'Conferences' or 'Conference Proceedings,' as well as listed by subject, and author in the appropriate indexes. |
| **Conference Coverage** | States in beginning of some issues that they "select significant items". Papers from conferences appear to be covered in highly selective fashion, usually only one or two papers from a conference. Several different CAB journals may cover the same Conference, selecting different papers from each conference for inclusion in the various journals. |
| **Abstracts Note** | All materials that are included are abstracted, including conferences and papers. |
| **Ordering Information** | Many issues include a form and detailed instructions for ordering photocopies of listed papers from CAB's Document Delivery Service. |
| **Electronic Equivalent** | Available online as CAB Abstracts, accessible as a subfile in the database from 1972- on BRS, Dialog (file 50) and STN. Available on CD-ROM 1984- through SilverPlatter. |
| **Notes** | The journals are prepared centrally, but indexed by each separate CAB Institute. |

## Forestry Abstracts

| | |
|---|---|
| **Producer** | C.A.B. International (Commonwealth Agricultural Bureaux)<br>845 North Park Ave.<br>Tucson, AZ 85719 |
| **Subjects** | Agriculture |
| **Publication Dates** | 1939- |
| **Access Points** | Conferences are not listed in 'Contents.'<br>Author, Subject, broad divisional by subjects on 'Contents' page. |
| **Conference Indexing** | Individual conference papers are listed in the annual subject index under the heading 'Conferences' or 'Conference Proceedings,' as well as listed by subject, and author in the appropriate indexes. |
| **Conference Coverage** | States in beginning of some issues that they "select significant items". Papers from conferences appear to be covered in highly selective fashion, usually only one or two papers from a conference. Several different CAB journals may cover the same Conference, selecting different papers from each conference for inclusion in the various journals. |
| **Abstracts Note** | All materials that are included are abstracted, including conferences and papers |
| **Ordering Information** | Many issues include a form for ordering photocopies of listed papers from CAB's Document Delivery Service, and detailed instructions. |
| **Electronic Equivalent** | Available online as CAB Abstracts, accessible as a subfile in the database from 1972- on BRS, Dialog (file 50) and STN. Available on CD-ROM 1984- through SilverPlatter. |
| **Notes** | The journals are prepared centrally, but indexed by each separate CAB Institute. |

## Food Science and Technology Abstracts (FSTA)

| | |
|---|---|
| **Producer** | International Food Information Service (IFIS)<br>Lane End House<br>Shinfield<br>Reading RG2 9BB England |
| **Subjects** | Food Industry and Trade |
| **Publication Dates** | 1969- |
| **Access Points** | Author and Subject Indexes. Monthly and Annual cumulations. |
| **Conference Indexing** | Conference as a whole is indexed by subject. "Conference proceedings" is one of the subject headings. Many, but not all, conferences also have the individual papers indexed by author and subject. If not, then the abstract for the conference lists all papers and authors presented. |
| **Conference Coverage** | Provides wide coverage and indexes monographs as well as conference proceedings issued as parts of journals. |
| **Abstracts Note** | Abstract for each entry; contains "details of new information." |
| **Ordering Information** | |
| **Electronic Equivalent** | Available online 1969- on BRS, Dialog (file 51) and ORBIT. Available on CD-ROM 1969- through SilverPlatter. |
| **Notes** | |

## Gas Abstracts

| | |
|---|---|
| **Producer** | Institute of Gas Technology<br>3424 So. State St.<br>Chicago, IL 60616 |
| **Subjects** | Energy |
| **Publication Dates** | 1945- |
| **Access Points** | Author, Conference author, Subject, Broad divisional by subjects in "contents" page. |
| **Conference Indexing** | Listed by author and subject. No special section for conference proceedings, no subject heading for "Conference" or "Conference Proceedings". |
| **Conference Coverage** | Most conferences covered selectively (one abstract for entire conference), some, especially those of interest locally, e.g. PSIG (Pipeline Simulation Interest Group) Annual Meetings are covered comprehensively. |
| **Abstracts Note** | All materials included are abstracted, including conferences and papers. |
| **Ordering Information** | Each issue includes a photocopy order form with detailed instructions. |
| **Electronic Equivalent** | Available online as IGT's "Gasline". |
| **Notes** | Worldwide coverage; English, Roman alphabet, Cyrillic group languages covered. |

## General Science Index

| | |
|---|---|
| **Producer** | H. W. Wilson Company<br>950 University Avenue<br>Bronx, New York 10452 |
| **Subjects** | Multidisciplinary |
| **Publication Dates** | June 1978- |
| **Access Points** | Subject index. Monthly publication with quarterly and annual cumulations. |
| **Conference Indexing** | Conferences are indexed under subject headings; however, GSI is only incidentally concerned with proceedings. Online searching through words such as "proceedings," "conference," or "symposium" will also provide access. |
| **Conference Coverage** | GSI does not cover many conferences. Journals such as the *American Zoologist* sometimes publish papers from symposia and these are indexed individually. |
| **Abstracts Note** | No abstracts are provided by this Wilson index. |
| **Ordering Information** | H. W. Wilson has no document delivery service. |
| **Electronic Equivalent** | Available online through WILSONLINE since May 1984. Available on CD-ROM 1984- through WilsonDisk, SilverPlatter and CD Plus Technologies. |
| **Notes** | Since GSI is directed toward a relatively unsophisticated audience, conference proceedings are not a primary concern. However, as of August 1, 1990, there are 148 included in the database. |

## Geographical Abstracts - Human Geography

| | |
|---|---|
| **Producer** | GEO Abstracts/Elsevier<br>Regency House, 34 Duke St.<br>NR3 3AP<br>Norwich<br>England |
| **Subjects** | Geography |
| **Publication Dates** | 1966- |
| **Access Points** | Author and general subject in each bi-monthly issue; Annual cumulated index to all sections by author and keywords from the titles (KWIC) |
| **Conference Indexing** | No special indication or handling of conference papers. Whole conference proceedings are not indexed if published monographically. |
| **Conference Coverage** | About 10% of the citations are to conference papers or proceedings. Most conference citations are to individual conference papers and/or whole proceedings which were published as journal articles. A few of the *Journals Abstracted in Geographical Abstracts* are society proceedings. |
| **Abstracts Note** | All citations have abstracts |
| **Ordering Information** | |
| **Electronic Equivalent** | Available 1980- as GEOBASE in Dialog (file 292) and ORBIT. Available on CD-ROM 1980- through SilverPlatter. |
| **Notes** | The sections of Geographical Abstracts that merged to form this title are:<br>C. Economic Geography<br>D. Social and Historical Geography<br>F. Regional and Community Planning |

## Geographical Abstracts - Physical Geography

| | |
|---|---|
| **Producer** | GEO Abstracts/Elsevier<br>Regency House, 34 Duke St.<br>NR3 3AP<br>Norwich<br>England |
| **Subjects** | Geography, Geosciences |
| **Publication Dates** | 1966- |
| **Access Points** | Author and general subject in each bi-monthly issue; Annual cumulated index to all sections by author and keywords from the titles (KWIC) |
| **Conference Indexing** | No special indication or handling of conference papers. Whole conference proceedings are not indexed if published monographically. |
| **Conference Coverage** | About 10% of the citations are to conference papers or proceedings. Most conference citations are to individual conference papers and/or whole proceedings which were published as journal articles. A few of the *Journals Abstracted in Geographical Abstracts* are society proceedings. |
| **Abstracts Note** | All citations have abstracts |
| **Ordering Information** | |
| **Electronic Equivalent** | Available 1980- as GEOBASE in Dialog (file 292) and ORBIT. Available on CD-ROM 1980- through SilverPlatter. |
| **Notes** | The sections of Geographical Abstracts that merged to form this title are:<br>A. Landforms and the Quaternary<br>B. Climatology and Hydrology<br>E. Sedimentology<br>G. Remote Sensing Photogrammetry and Cartography |

## Geological Abstracts

| | |
|---|---|
| **Producer** | GEO Abstracts/Elsevier<br>Regency House, 34 Duke St.<br>NR3 3AP<br>Norwich<br>England |
| **Subjects** | Geosciences |
| **Publication Dates** | Varies with section:<br>Economic Geology: 1986-<br>Geophysics and Tectonics: 1977-<br>Palaeontology and Stratigraphy: 1986-<br>Sedimentary Geology: 1986- |
| **Access Points** | Author and general subject |
| **Conference Indexing** | No special indication or handling of conference papers. Whole conference proceedings are not indexed if published monographically. |
| **Conference Coverage** | About 10% of the citations are to conference papers or proceedings. Most conference citations are to individual conference papers and/or whole proceedings which were published as journal articles. A few of the *Journals Abstracted in Geographical Abstracts* are society proceedings. |
| **Abstracts Note** | All citations have abstracts |
| **Ordering Information** | |
| **Electronic Equivalent** | Available 1980- as GEOBASE in Dialog (file 292) and ORBIT. Available on CD-ROM 1980- through SilverPlatter. |
| **Notes** | |

## Horticultural Abstracts

| | |
|---|---|
| **Producer** | C.A.B. International (Commonwealth Agricultural Bureaux)<br>845 North Park Ave.<br>Tucson, AZ, 85719 |
| **Subjects** | Agriculture, Botany |
| **Publication Dates** | 1931- |
| **Access Points** | Conferences listed in 'Contents.'<br>Author, Subject, broad divisional by subjects on 'Contents' page. |
| **Conference Indexing** | Conferences are listed in 'Contents' and individual papers are listed in the annual subject index under the heading 'Conferences' or 'Conference Proceedings,' as well as listed by subject, and author in the appropriate indexes. |
| **Conference Coverage** | States in beginning of some issues that they "select significant items". Papers from conferences appear to be covered in highly selective fashion, usually only one or two papers from a conference. Several different CAB journals may cover the same Conference, selecting different papers from each conference for inclusion in the various journals. |
| **Abstracts Note** | All materials that are included are abstracted, including conferences and papers. |
| **Ordering Information** | Many issues include a form for ordering photocopies of listed papers from CAB's Document Delivery Service, and detailed instructions. |
| **Electronic Equivalent** | Available online as CAB Abstracts, accessible as a subfile in the database from 1972- on BRS, Dialog (file 50) and STN. Available on CD-ROM 1984- through SilverPlatter. |
| **Notes** | The journals are prepared centrally, but indexed by each separate CAB Institute. |

## Index of Conference Proceedings Received

| | |
|---|---|
| **Producer** | British Library Document Supply Centre<br>Boston Spa<br>Wetherby, W. Yorkshire<br>LS23 7BQ<br>England |
| **Subjects** | Multidisciplinary |
| **Publication Dates** | 1964-; however, selected coverage back to the 19th Century |
| **Access Points** | KWIC and Subject |
| **Conference Indexing** | Individual papers are not indexed. |
| **Conference Coverage** | Worldwide coverage of the monographic conference proceedings literature. |
| **Abstracts Note** | No abstracts |
| **Ordering Information** | All items are available on loan to organizations in the UK registered as users of the Lending Division. Photocopies of extracts from these publications can be supplied to any organization both in the UK and overseas through the Photocopy Service, subject to the terms of the Copyright Act. |
| **Electronic Equivalent** | |
| **Notes** | |

## Index to Scientific and Technical Proceedings

| | |
|---|---|
| **Producer** | ISI (Institute for Scientific Information)<br>3501 Market St.<br>Philadelphia, PA 19104 |
| **Subjects** | Multidisciplinary |
| **Publication Dates** | 1978- |
| **Access Points** | Author/Editor, Broad Subject Category, Contents of Proceedings, Sponsor, Meeting, Location, Permuterm Subject, Corporate (both geographic and organization name indexes) |
| **Conference Indexing** | See above Access Points list. The Contents of Proceedings index lists all papers under the full record for the proceedings. |
| **Conference Coverage** | Published proceedings only, regardless of format of publication. Only significant proceedings are selected for coverage. Societies and publishers are contacted for material to index. Worldwide coverage. A list of the books in series that are covered is included. Approximately 5000 proceedings a year are indexed. |
| **Abstracts Note** | No abstracts |
| **Ordering Information** | Some papers available through ISI's document delivery system; these are labeled as such. |
| **Electronic Equivalent** | Available online 1982- through ISI. |
| **Notes** | |

## Index Veterinarius

| | |
|---|---|
| **Producer** | C.A.B. International (Commonwealth Agricultural Bureaux)<br>845 North Park Ave.<br>Tucson, AZ, 85719 |
| **Subjects** | Veterinary Science |
| **Publication Dates** | 1933- |
| **Access Points** | Author, Subject, Broad subject divisions in 'Contents' pages |
| **Conference Indexing** | Conferences are listed in 'Contents' and individual papers are listed in the annual subject index under the heading 'Conferences' or 'Conference Proceedings,' as well as listed by subject, and author in the appropriate indexes. |
| **Conference Coverage** | States in beginning of some issues that they select "significant items". Papers from conferences appear to be covered in highly selective fashion, usually only one or two papers from a conference. Several different CAB journals may cover the same Conference, selecting different papers from each conference for inclusion in the various journals. |
| **Abstracts Note** | Abstracts available for some citations to individual papers but not for whole proceedings. |
| **Ordering Information** | Many issues include a form and detailed instructions for ordering photocopies of listed papers from CAB's Document Delivery Service. |
| **Electronic Equivalent** | Available online as CAB International, accessible as a subfile in the database from 1972- on BRS, Dialog (file 50) and STN. Available on CD-ROM 1984- through SilverPlatter. |
| **Notes** | The journals are prepared centrally, but indexed by each separate CAB Institute. |

## INIS Atomindex

| | |
|---|---|
| **Producer** | International Atomic Energy Agency<br>dist. by<br>UNIPUB<br>4611-F Assembly Drive<br>Lanham, MD 20706 |
| **Subjects** | Nuclear Science, Energy |
| **Publication Dates** | 1970- |
| **Access Points** | Personal Author, Title and Subject |
| **Conference Indexing** | Provides one record for the whole conference and separate records for selected papers. Meeting date and year are searchable fields in the online file. 'Conference' is a searchable document type in the online file. 'Proceedings' is an index term for records for a whole conference. |
| **Conference Coverage** | About 28% of the citations in the online file are to conference papers. |
| **Abstracts Note** | Abstracts are available for most citations. |
| **Ordering Information** | |
| **Electronic Equivalent** | Available online 1974- as ENERGY on STN. Available on CD-ROM 1976- through SilverPlatter. |
| **Notes** | Incorporates Nuclear Science Abstracts. |

## Mathematical Reviews

| | |
|---|---|
| **Producer** | American Mathematical Society<br>P.O. Box 6248<br>Providence, RI 02940 |
| **Subjects** | Mathematics |
| **Publication Dates** | 1940- |
| **Access Points** | Author, Broad subject classification in monthly issues; author and subject in the annual indexes. |
| **Conference Indexing** | The subheading number 06, "Proceedings, conferences, collections, etc." is used under the broad subject headings for whole proceedings. Records for whole conference proceedings include a list of the papers in the proceedings, but no abstract or review. Not every paper in a proceedings has an individual entry and abstract or review. In the annual index, a section is devoted to listing papers by author that are published in conference proceedings and collections of papers. These entries refer to the full citation for the individual papers. In the online file, 'proceedings paper' and 'proceedings' are searchable as document type. The subheading "Proceedings, conferences, collections, etc." is applied also. |
| **Conference Coverage** | Worldwide coverage of conferences, no limitation on languages. |
| **Abstracts Note** | Signed, in-depth reviews or author's abstracts are available for most papers. |
| **Ordering Information** | Document ordering is available from the AMS through the MATHDOC service. |
| **Electronic Equivalent** | Available online 1940 - as MATHSCI on Dialog (file 239) and STN. Available on CD-ROM 1940- through SilverPlatter. |
| **Notes** | |

## Mechanical Engineering Abstracts

| | |
|---|---|
| **Producer** | Cambridge Scientific Abstracts<br>7200 Wisconsin Avenue<br>Bethesda, MD 20814 |
| **Subjects** | Engineering, Energy |
| **Publication Dates** | 1967- |
| **Access Points** | Personal Author, Title, Subject |
| **Conference Indexing** | Provides separate records for each paper in the conference. Conference title, conference location, conference organizer and conference year are searchable fields in the online version. |
| **Conference Coverage** | Conferences falling into subject areas covered by the index are included. For example, all relevant conferences of the IEEE are included. |
| **Abstracts Note** | Abstracts are included for all materials indexed from 1982 to present. |
| **Ordering Information** | |
| **Electronic Equivalent** | Available online 1973- as ISMEC on Dialog (file 14) and STN. Available on CD-ROM 1973- through SilverPlatter. |
| **Notes** | List of descriptors is available from Cambridge Scientific Abstracts. |

## Metals Abstracts and Metals Abstracts Index

| | |
|---|---|
| **Producer** | Materials Information, a joint service of:<br>The Institute of Materials and<br>ASM International<br>Materials Park, Ohio 44073 |
| **Subjects** | Metals, Metallurgy, Materials Science |
| **Publication Dates** | 1968-; continues other titles covering metallurgy from 1934. |
| **Access Points** | Personal and Corporate Author, Subject; monthly and annual cumulations. |
| **Conference Indexing** | Provides one record for the whole conference and separate records for selected papers. Majority are selectively indexed. List of "Special Publications" includes conferences and books giving title, date, place, collation, language, price, ISBN. Abstracts for these appear elsewhere in the issue. Conference title and conference year are searchable fields in the online version. |
| **Conference Coverage** | Conferences falling into subject areas covered by the index are included. For example, all conferences of the ASTM, AIME, & Metal Research Society are included. Relevant meetings of other groups, such as the ACS and the Physical Society, are covered. |
| **Abstracts Note** | Abstracts are included for all materials indexed. For translated materials the abstract appears only the first time the item is included. When the translation appears later, a reference to the original abstract is given. |
| **Ordering Information** | Materials Information provides document delivery service from the USA and UK offices. Items not available: patents, dissertations, NTIS documents, some translation journal articles. |
| **Electronic Equivalent** | Available online 1966- as METADEX on Dialog (file 14), Orbit, STN, ESA-IRS, CEDOCAR, CISTI, CAN/OLE, Data-Star and FIZ Technik. Available on CD-ROM 1985- through Dialog OnDisc. |
| **Notes** | Controlled subject vocabulary in *Thesaurus of Metallurgical Terms.* |

## Meteorological and Geoastrophysical Abstracts

| | |
|---|---|
| **Producer** | American Meteorological Society<br>45 Beacon St.<br>Boston, MA 02108 |
| **Subjects** | Meteorology, Astronomy, Geosciences |
| **Publication Dates** | 1950- |
| **Access Points** | Author, general subject, geographic |
| **Conference Indexing** | No special annotations for conferences; no conference proceedings listed with journals indexed. |
| **Conference Coverage** | Most conference citations are to individual conference papers and/or whole proceedings which were published as journal articles. Worldwide coverage. |
| **Abstracts Note** | All citations have abstracts |
| **Ordering Information** | |
| **Electronic Equivalent** | Available online 1970- on Dialog (file 29). |
| **Notes** | |

## Mineralogical Abstracts

**Producer** Mineralogical Society
41 Queen's Gate
London SW7 5HR

**Subjects** Geosciences

**Publication Dates** 1920-

**Access Points** Author and general subject

**Conference Indexing** No special annotations for conferences.

**Conference Coverage** About 10% of the citations are to conference papers or proceedings. Most conference citations are to individual conference papers and/or whole proceedings which were published as journal articles. A few of the Journals *Abstracted in Mineralogical Abstracts* are society proceedings. Worldwide coverage.

**Abstracts Note** All citations have abstracts

**Ordering Information**

**Electronic Equivalent** Available online 1982- as GEOBASE on Dialog (file 292) and ORBIT. Available on CD-ROM 1980- through SilverPlatter.

**Notes**

## NTIS

| | |
|---|---|
| **Producer** | National Technical Information Service<br>U.S. Dept. of Commerce<br>5285 Port Royal Road<br>Springfield, VA 22161 |
| **Subjects** | Multidisciplinary |
| **Publication Dates** | 1964- |
| **Access Points** | Author, Title and Subject |
| **Conference Indexing** | Document Type 'Conference Proceeding' is searchable. Both individual papers and full conference proceedings are indexed. |
| **Conference Coverage** | International coverage. |
| **Abstracts Note** | Abstracts are available for some papers and proceedings. |
| **Ordering Information** | NTIS provides a document delivery service. |
| **Electronic Equivalent** | Available online as NTIS through Dialog (file 6) and STN. Available on CD-ROM 1983- through SilverPlatter. |
| **Notes** | NTIS incorporates all of Government Reports Announcements and Index |

## Nutrition Abstracts and Reviews.
## Section A: Human and Experimental
## Section B: Livestock Feeds and Feeding

| | |
|---|---|
| **Producer** | C.A.B. International (Commonwealth Agricultural Bureaux)<br>845 North Park Ave.<br>Tucson, AZ, 85719 |
| **Subjects** | Nutrition |
| **Publication Dates** | 1931- (split in 1977 into Series A & B) |
| **Access Points** | Author, Subject, broad subject divisions on 'Contents' page. |
| **Conference Indexing** | Conferences are listed in 'Contents' and individual papers are listed in the annual subject index under the heading 'Conferences' or 'Conference Proceedings,' as well as listed by subject, and author in the appropriate indexes. |
| **Conference Coverage** | States in beginning of some issues that they "select significant terms". Papers from conferences appear to be covered in highly selective fashion, usually only one or two papers from a conference. Several different CAB journals may cover the same Conference, selecting different papers from each conference for inclusion in the various journals. |
| **Abstracts Note** | All materials that are included are abstracted, including conference proceedings and papers |
| **Ordering Information** | Many issues include a form and detailed instructions for ordering photocopies of listed papers from C.A.B.'s Document Delivery Service. |
| **Electronic Equivalent** | Available online as CAB Abstracts, accessible as a subfile in the database from 1972- on BRS, Dialog (file 50) and STN. Available on CD-ROM 1984- through SilverPlatter. |
| **Notes** | The journals are prepared centrally, but indexed by each separate CAB Institute. |

## Oceanic Abstracts

| | |
|---|---|
| **Producer** | Cambridge Scientific Abstracts<br>7200 Wisconsin Avenue<br>Bethesda, MD 20814 |
| **Subjects** | Oceanography, Marine Biology, Environmental Studies |
| **Publication Dates** | 1964- |
| **Access Points** | Author, Subject, Taxonomic, Geographic |
| **Conference Indexing** | Individual papers are cited in this index instead of using one record for the whole conference. There is a separate section of abstracts of conference papers. In the online database, document type is a searchable field. |
| **Conference Coverage** | Coverage of conferences is about 22% of citations in this index. English is the predominant language but coverage is international. |
| **Abstracts Note** | Almost all citations have abstracts. |
| **Ordering Information** | Cambridge Scientific Abstracts/ASFA suggests the British Library Lending Division, Information on Demand (IOD), the National Agricultural Library, and the National Technical Information Service. |
| **Electronic Equivalent** | Available online 1964- on STN and Dialog (file 28). Available on CD-ROM from Cambridge Scientific Abstracts. |
| **Notes** | |

## PASCAL

| | |
|---|---|
| **Producer** | ICNRS/INIST<br>2, Allee du Parc de Brabois<br>54514 Vandpeuvre-les-Nancy CEDEX<br>France |
| **Subjects** | Multidisciplinary |
| **Publication Dates** | 1973- |
| **Access Points** | Author, Title, Descriptors in several languages |
| **Conference Indexing** | There is one record for the whole conference proceedings if published as a monograph. Generally, there is a record for each individual paper. Document type "conference proceedings" is searchable in the online database. Conference title, location and year are searchable fields in the online file. |
| **Conference Coverage** | Conference records make up less than 7% of the file. |
| **Abstracts Note** | Abstracts are available for some records. |
| **Ordering Information** | |
| **Electronic Equivalent** | Available online 1969- as PASCAL on Dialog (file 144). |
| **Notes** | Formerly Bulletin Signaletique. |

## Physics Abstracts

| | |
|---|---|
| **Producer** | IEE (Institution of Electrical Engineers)<br>P.O. Box 1331<br>Piscataway, NJ 08855 |
| **Subjects** | Physics, Engineering |
| **Publication Dates** | 1903- |
| **Access Points** | Author, Subject from classification scheme, Keyword subject, multi-year cumulations of the indexes |
| **Conference Indexing** | There is one record for the whole conference proceedings if published as a monograph, plus a record for each individual paper. Document types "conference paper", "conference proceedings" and "conference article" are searchable fields in the online database. |
| **Conference Coverage** | About 21% of the citations are to conferences proceedings or papers. The IEE Proceedings series and the IEEE Transactions series are indexed. Conferences published in serial form that are routinely indexed are listed in the *List of Journals and other Serials Sources.* |
| **Abstracts Note** | Abstracts are available for all records. The abstract for a whole conference proceedings lists the topics covered in individual papers, which each has its own abstract. |
| **Ordering Information** | |
| **Electronic Equivalent** | Available online 1969- as INSPEC on BRS, Dialog (file 2) and STN. Available on CD-ROM 1989- through UMI Ondisc. |
| **Notes** | |

## Physics Briefs

| | |
|---|---|
| **Producer** | FIZ Karlsruhe<br>P.O. Box 2465<br>D-7500 Karlsruhe 1<br>Germany<br>and<br>American Institute of Physics<br>335 East 45th St.<br>New York, NY 10017 |
| **Subjects** | Physics, Astronomy, Astrophysics |
| **Publication Dates** | 1845-1994 |
| **Access Points** | Author, Subject |
| **Conference Indexing** | There is one record for the whole conference proceedings if published as a monograph, plus a record for each individual paper. Document type "conference" is a searchable field in the online database. |
| **Conference Coverage** | 28% of the citations are to conference papers and proceedings. Coverage is worldwide. Proceedings published in serial form that are routinely indexed are listed in the *List of Journals and Serial Publications*. |
| **Abstracts Note** | Most citations have short abstracts. |
| **Ordering Information** | |
| **Electronic Equivalent** | Was available online 1979-1994 as PHYS on STN. Records from 1995- are also available online through the INSPEC database on STN and Dialog (file 2). |
| **Notes** | |

## Plant Breeding Abstracts

| | |
|---|---|
| **Producer** | C.A.B. International (Commonwealth Agricultural Bureaux)Farnham Royal<br>845 North Park Ave.Slough<br>Tucson, AZ 85719 |
| **Subjects** | Agriculture |
| **Publication Dates** | 1930- |
| **Access Points** | Author, Subject, Broad subject divisions in 'Contents' pages |
| **Conference Indexing** | Conferences are listed in 'Contents' and individual papers are listed in the subject index under the heading 'Conferences' or 'Conference Proceedings,' as well as listed by subject, and author in the appropriate indexes. |
| **Conference Coverage** | States in beginning of some issues that they "select significant items". Papers from conferences appear to be covered in highly selective fashion, usually only one or two papers from a conference. Several different CAB journals may cover the same Conference, selecting different papers from each conference for inclusion in the various journals. |
| **Abstracts Note** | All materials that are included are abstracted, including conferences and conference papers |
| **Ordering Information** | Many issues include a form and detailed instructions for ordering photocopies of listed papers from C.A.B.'s Document Delivery Service |
| **Electronic Equivalent** | Available online as CAB Abstracts, accessible as a subfile in the database from 1972- on BRS, Dialog (file 50) and STN. Available on CD-ROM 1984- through SilverPlatter. |
| **Notes** | The journals are prepared centrally, but indexed by each separate CAB Institute. |

## Pollution Abstracts

| | |
|---|---|
| **Producer** | Cambridge Scientific Abstracts<br>7200 Wisconsin Avenue<br>Bethesda, MD 20814 |
| **Subjects** | Pollution, Environmental Sciences |
| **Publication Dates** | 1970- |
| **Access Points** | Author, Subject |
| **Conference Indexing** | Individual papers are cited in this index instead of using one record for the whole conference. In the online database, document type is a searchable field. |
| **Conference Coverage** | Coverage of conferences is about 22% of citations in this index. English is the predominant language but coverage is international. |
| **Abstracts Note** | Almost all citations have abstracts. |
| **Ordering Information** | Cambridge Scientific Abstracts/ASFA suggests the British Library Lending Division, Information on Demand (IOD), the National Agricultural Library, and the National Technical Information Service. |
| **Electronic Equivalent** | Available online 1970- on BRS and Dialog (file 41). Available on CD-ROM through SilverPlatter as part of PolTox1. |

**Notes**

## Proceedings in Print

| | |
|---|---|
| **Producer** | Proceedings in Print, Inc.<br>Box 369<br>Halifax, MA 02338 |
| **Subjects** | Multidisciplinary |
| **Publication Dates** | 1964- |
| **Access Points** | Title of the proceedings or conference, Corporate author, Sponsoring agency, Editor, Keyword or Subject Heading |
| **Conference Indexing** | All citations are to full conference proceedings. |
| **Conference Coverage** | International coverage. The conference must have been published to be included, so there are no citations to "in press" materials. However, there are some citations to conferences for which it was known that the proceedings would never be published. |
| **Abstracts Note** | No abstracts. |
| **Ordering Information** | Ordering information is given for each item if it is known. |
| **Electronic Equivalent** | |
| **Notes** | |

## Rapra Abstracts

| | |
|---|---|
| **Producer** | Rapra Technology Ltd<br>Shawbury, Shrewsbury<br>Shropshire SY4 4NR<br>United Kingdom |
| **Subjects** | Rubber, Plastics, Engineering |
| **Publication Dates** | 1923- |
| **Access Points** | Personal Author, Title, Subject |
| **Conference Indexing** | Provides one record for the conference as a whole as well as separate records for each paper in the conference. 'Conferences' is a subject heading for records describing a whole conference proceedings or set of papers. Conference title is a searchable fields in the online version. |
| **Conference Coverage** | Conferences falling into subject areas covered by the index are included. |
| **Abstracts Note** | Abstracts are included for all materials indexed. |
| **Ordering Information** | |
| **Electronic Equivalent** | Available online 1972- as RAPRA on Dialog (file 323) and STN. |
| **Notes** | Rapra Keyterm Thesaurus is available from the producer. Online version includes Adhesives Abstracts and Rapra New Tradenames. |

**Review of Agricultural Entomology**

| | |
|---|---|
| **Producer** | C.A.B. International (Commonwealth Agricultural Bureaux)Farnham Royal<br>845 North Park Ave.Slough<br>Tucson, AZ 85719 |
| **Subjects** | Entomology |
| **Publication Dates** | 1913- |
| **Access Points** | Author, Subject, Broad subject divisions in 'Contents' pages |
| **Conference Indexing** | Conferences are listed in 'Contents' and individual papers are listed in the annual subject index under the heading 'Conferences' or 'Conference Proceedings,' as well as listed by subject, and author in the appropriate indexes. |
| **Conference Coverage** | States in beginning of some issues that they "select significant items". Papers from conferences are covered in highly selective fashion, usually only one or two papers from a conference. Several different CAB journals may cover the same Conference, selecting different papers from each conference for inclusion in the various journals. |
| **Abstracts Note** | All materials that are included are abstracted, including conferences and papers. |
| **Ordering Information** | Many issues include a form and detailed instructions for ordering photocopies of listed papers from CAB's Document Delivery Service. |
| **Electronic Equivalent** | Available online as CAB Abstracts, accessible as a subfile in the database from 1972- on BRS, Dialog (file 50) and STN. Available on CD-ROM 1984- through SilverPlatter. |
| **Notes** | The journals are prepared centrally, but indexed by each separate CAB Institute. |

## Review of Plant Pathology

| | |
|---|---|
| **Producer** | C.A.B. International (Commonwealth Agricultural Bureaux)<br>845 North Park Ave.<br>Tucson, AZ, 85719 |
| **Subjects** | Botany |
| **Publication Dates** | 1921- |
| **Access Points** | Author, Subject, Broad subject divisions in 'Contents' pages |
| **Conference Indexing** | Conferences are listed in 'Contents' and individual papers are listed in the annual subject index under the heading 'Conferences' or 'Conference Proceedings,' as well as listed by subject, and author in the appropriate indexes. |
| **Conference Coverage** | States in beginning of some issues that they "select significant items". Papers from conferences appear to be covered in highly selective fashion, usually only one or two papers from a conference. Several different CAB journals may cover the same Conference, selecting different papers from each conference for inclusion in the various journals. |
| **Abstracts Note** | All materials that are included are abstracted, including conferences and papers |
| **Ordering Information** | Many issues include a form and detailed instructions for ordering photocopies of listed papers from CAB's Document Delivery Service. |
| **Electronic Equivalent** | Available online as CAB Abstracts, accessible as a subfile in the database from 1972- on BRS, Dialog (file 50) and STN. Available on CD-ROM 1984- through SilverPlatter. |
| **Notes** | The journals are prepared centrally, but indexed by each separate C.A.B. Institute. |

## Robotics Abstracts

| | |
|---|---|
| **Producer** | Bowker A & I Publishing<br>R.R.Bowker<br>245 West 17th St.<br>New York, New York 10011 |
| **Subjects** | Computer Science, Engineering |
| **Publication Dates** | 1973- |
| **Access Points** | Author, Subject, Industry, Source |
| **Conference Indexing** | Document types such as 'conference proceedings' or 'conference paper' are indicated on each record, but not indexed as such. A separate section "Conferences and Events" provides dates and locations of selected conferences. In the online databases, document type is a searchable field. |
| **Conference Coverage** | Core conference sources are scanned. Conference papers make up 30% in the energy and environment indexes to 50% in the computer indexes of the items in these indexes. Only English language materials are included. |
| **Abstracts Note** | All citations have abstracts. |
| **Ordering Information** | Document Retrieval Service makes documents available in hard copy or microfiche. |
| **Electronic Equivalent** | Was 1985- in the Supertech file on Dialog; no longer available. |
| **Notes** | Supertech includes records from two discontinued titles, Telecommunications Abstracts and Telegen Abstracts. Supertech in Dialog includes Artificial Intelligence Abstracts, CAD/CAM Abstracts, Robotics Abstracts |

## Science Citation Index

| | |
|---|---|
| **Producer** | ISI (Institute for Scientific Information)<br>3501 Market St.<br>Philadelphia, PA 19104 |
| **Subjects** | Multidisciplinary |
| **Publication Dates** | 1961- |
| **Access Points** | Author, Permuterm subject, Corporate, Citation |
| **Conference Indexing** | There is no particular indexing for conferences. If one of the terms for conferences were in a title that would be found in the permuterm subject index. |
| **Conference Coverage** | No attempt is made to cover conferences; ISI's publication *Index to Scientific and Technical Proceedings* serves that function. |
| **Abstracts Note** | Abstracts available since 1990. |
| **Ordering Information** | ISI provides a document supply service. |
| **Electronic Equivalent** | Available 1974- as SCISEARCH on Dialog (file 34) and STN. Available on CD-ROM 1980- through ISI. |
| **Notes** | |

## Selected Water Resources Abstracts (SWRA)

| | |
|---|---|
| **Producer** | Water Resources Scientific Information Center<br>U.S. Geological Survey<br>U.S. Dept. Interior<br>425 National Center<br>Reston, VA 22092 |
| **Subjects** | Water Resources, Pollution |
| **Publication Dates** | 1968- |
| **Access Points** | 10 general subject categories; also subject, author, organization, and accession number indexes |
| **Conference Indexing** | No special annotation for conferences. |
| **Conference Coverage** | |
| **Abstracts Note** | Abstracts available for all citations. |
| **Ordering Information** | NTIS |
| **Electronic Equivalent** | Available 1968- as Water Resources Abstracts on Dialog (file 117). Available on CD-ROM 1967- through SilverPlatter and Cambridge Scientific Abstracts. |
| **Notes** | |

## Soils and Fertilizers

| | |
|---|---|
| **Producer** | C.A.B. International (Commonwealth Agricultural Bureaux)<br>845 North Park Ave.<br>Tucson, AZ 85719 |
| **Subjects** | Soil Science |
| **Publication Dates** | 1938- |
| **Access Points** | Author, Subject, Broad subject divisions in 'Contents' pages. |
| **Conference Indexing** | Conferences are listed in 'Contents' and individual papers are listed in the annual subject index under the heading 'Conferences' or 'Conference Proceedings,' as well as listed by subject, and author in the appropriate indexes. |
| **Conference Coverage** | States in beginning of some issues that "selects significant items". Papers from conferences are covered in highly selective fashion, usually only one or two papers from a conference. Several different CAB journals may cover the same conference, selecting different papers from each conference for inclusion in the various journals. |
| **Abstracts Note** | All materials that are included are abstracted, including conference proceedings and papers |
| **Ordering Information** | Many issues include a form and detailed instructions for ordering photocopies of listed papers from CAB's Document Delivery Service. |
| **Electronic Equivalent** | Available online as CAB International, accessible as a subfile from 1972- in BRS, Dialog (file 50) and STN. Available on CD-ROM 1984- through SilverPlatter. |
| **Notes** | The journals are prepared centrally, but indexed by each separate CAB Institute. |

## Solid State and Superconductivity Abstracts

**Producer**
Cambridge Scientific Abstracts
7200 Wisconsin Avenue
Bethesda, MD 20814

**Subjects**
Physics

**Publication Dates**
1960-

**Access Points**
Personal Author, Title, Subject, Abstract

**Conference Indexing**
Provides separate records for each paper in the conference, and a record for the conference proceedings as a whole. Conference title, conference location, and conference date and year are searchable fields in the online version.

**Conference Coverage**
Conferences falling into subject areas covered by the index are included. For example, all relevant conferences of the IEEE are included.

**Abstracts Note**
Abstracts are included for all materials indexed.

**Ordering Information**

**Electronic Equivalent**
Available online 1981- as SOLIDSTATE on STN.

**Notes**

## Weed Abstracts

| | |
|---|---|
| **Producer** | C.A.B. International (Commonwealth Agricultural Bureaux)<br>845 North Park Ave.<br>Tucson, AZ, 85719 |
| **Subjects** | Agriculture |
| **Publication Dates** | 1951 |
| **Access Points** | Author, Subject, Broad subject divisions in 'Contents' pages |
| **Conference Indexing** | Conferences are listed in 'Contents' and individual papers are listed in the subject index under the heading 'Conferences' or 'Conference proceedings', as well as listed by subject, and author in the appropriate indexes. |
| **Conference Coverage** | States in beginning of some issues that they "select significant items". Papers from conferences appear to be covered in highly selective fashion, usually only one or two papers from a conference. Several different CAB journals may cover the same Conference, selecting different papers from each conference for inclusion in the various journals. |
| **Abstracts Note** | All materials that are included are abstracted, including conferences and conference papers |
| **Ordering Information** | Many issues include a form and detailed instructions for ordering photocopies of listed papers from CAB's Document Delivery Service. |
| **Electronic Equivalent** | Available online as CAB Abstracts, accessible as a subfile in the database from 1972- on BRS, Dialog (file 50) and STN. Available on CD-ROM 1984- through SilverPlatter. |
| **Notes** | The journals are prepared centrally, but indexed by each separate CAB Institute. |

## Wildlife Review

| | |
|---|---|
| **Producer** | Fish and Wildlife Service (dist. USGPO)<br>1025 Pennock Pl.<br>Fort Collins, CO 80524 |
| **Subjects** | Wildlife |
| **Publication Dates** | 1935- |
| **Access Points** | Author, Subject, Systematic, Geographic |
| **Conference Indexing** | No special annotations for conferences. |
| **Conference Coverage** | Covers some conferences. |
| **Abstracts Note** | Less than 10% of the citations have abstracts. |
| **Ordering Information** | |
| **Electronic Equivalent** | Available on CD-ROM 1971- from NISC as part of Wildlife Review/Fisheries Review. |
| **Notes** | |

## Zentralblatt fur Mathematik und ihre Grenzegebiete /Mathematics Abstracts

| | |
|---|---|
| **Producer** | Springer-Verlag<br>44 Hartz Way<br>Secaucus, NJ 07096<br><br>Co-producer:<br>Heidelberger Akademie der Wissenschaften |
| **Subjects** | Mathematics |
| **Publication Dates** | 1931- |
| **Access Points** | Author, Title of paper, Subject, Mathematics Subject Classification codes |
| **Conference Indexing** | No special indexing for conference information. There are separate entries for full publications as well as for each paper within a publication. |
| **Conference Coverage** | About 13% of the citations are to conference papers. |
| **Abstracts Note** | Abstracts are available for all records since 1984. About half the abstracts are by reviewers, and comment on the quality of the material. |
| **Ordering Information** | |
| **Electronic Equivalent** | Available online 1972- as the MATH file on STN.<br>Available on CD-ROM from UMI. |
| **Notes** | Mathematics Subject Classification is available from producer. |

## Zoological Record

| | |
|---|---|
| **Producer** | BIOSIS<br>2100 Arch St.<br>Philadelphia, PA 19103-1399 |
| **Subjects** | Biology, Veterinary Science |
| **Publication Dates** | 1864- |
| **Access Points** | Author, Subject, Geographical, Paleontological, Systematic, or Biosystematic |
| **Conference Indexing** | An entry is made in the Subject index using the term 'Meeting'. Zoological Record provides one record for the whole conference and separate records for specific papers. |
| **Conference Coverage** | Papers are included if within the subject scope of Zoological Record, i.e. any aspect of animal biology, but there is an emphasis on natural behavior and systematics. There are no language or geographic restrictions. |
| **Abstracts Note** | No abstracts |
| **Ordering Information** | BIOSIS, U.K./Zoological Record suggest the British Library Document Supply Centre, British Museum of Natural History, and the Zoological Society of London. |
| **Electronic Equivalent** | Available online 1978- as Zoological Record on BRS and Dialog (file 185). Available on CD-ROM 1978- through SilverPlatter. |
| **Notes** | Zoological Record is published jointly by BIOSIS and the Zoological Society of London |

## 8. Title and Subject Index to List of Sources

AGRICOLA... 22
Agricultural Engineering Abstracts... 13
Agriculture... 13, 14, 15, 17, 22, 23, 24, 29, 37, 38, 45, 61, 72
Animal Breeding Abstracts... 14
Apicultural Abstracts...15
Applied Science and Technology Index... 16
AQUASCI... 17
Aquatic Sciences & Fisheries Abstracts... 17
Artificial Intelligence Abstracts... 18
Astronomy... 19, 53, 60
Astronomy and Astrophysics Abstracts... 19
Astrophysics... 19, 53, 60
Bees... 15
Bibliography and Index of Geology... 20
Bibliography of Agriculture... 22
Biochemistry... 23, 26
Biological Abstracts/RRM... 23
Biological and Agricultural Index... 24
Biology... 17, 23, 24, 35, 66, 75
BIOSIS... 23
Botany... 45, 66
Bulletin Signaletique... 58
CAB Abstracts... 13, 14, 15, 29, 37, 38, 45, 48, 56, 61, 65, 66, 70, 72
CA Search... 26
CAD/CAM Abstracts... 25
Chemical Abstracts... 26
Chemical Engineering... 25
Chemistry... 26
COMPENDEX... 34
Computer and Control Abstracts... 27
Computer Science... 18, 27, 67
Conference Papers Index... 28
CONFSCI... 28
Dairy Science Abstracts... 29
Directory of Published Proceedings. Series PCE... 30
Directory of Published Proceedings. Series SEMT... 31
Ecological Abstracts... 32
Ecology... 30, 32
ENERGY... 49
Energy... 33, 40, 49, 51
Energy Information Abstracts... 33
Energyline... 33
Engineering... 16, 34, 51, 59, 64, 67
Engineering Index... 34
Entomology... 15, 35, 65
Entomology Abstracts... 35
Enviroline... 36
Environment Abstracts... 36
Environmental Studies... 17, 32, 36, 55, 57

## Title and Subject Index to List of Sources, cont.

Field Crop Abstracts... 37
Fish and Fisheries... 17
Food Industry and Trade... 39
Food Science and Technology Abstracts (FSTA)... 39
Forestry Abstracts... 38
Gas Abstracts... 40
General Science Index... 41
GEOBASE... 32, 42, 43, 44, 54
Geographical Abstracts - Human Geography... 42
Geographical Abstracts - Physical Geography... 43
Geography... 42, 43
Geological Abstracts... 44
GeoRef... 20
Geosciences... 20, 43, 44, 53, 54
Government Reports Announcements and Index... 55
Horticultural Abstracts... 45
Index of Conference Proceedings Received... 46
Index to Scientific and Technical Proceedings... 47
Index Veterinarius... 48
INIS Atomindex... 49
INSPEC... 19, 27, 59, 60
ISMEC... 51
Life Sciences Collection... 35
Marine Biology... 57
Materials Science... 16, 26, 52
Mathematical Reviews... 50
Mathematics... 50, 74
MATHSCI... 50
Mechanical Engineering Abstracts... 51
METADEX... 52
Metal Abstracts and Metals Abstracts Index... 52
Metallurgy... 52
Metals... 52
Meteorological and Geoastrophysical Abstracts... 53
Meteorology... 53
Mineralogical Abstracts... 54
Multidisciplinary... 28, 31, 41, 46, 47, 55, 58, 63, 68
NTIS... 55
Nuclear Science... 49
Nutrition... 56
Nutrition Abstracts and Reviews... 56
Oceanic Abstracts... 57
Oceanography... 57
PASCAL... 58
PHYS... 19, 60
Physics... 59, 60, 71
Physics Abstracts... 59
Physics Briefs... 60
Plant Breeding Abstracts... 61
Plastics... 64

## Title and Subject Index to List of Sources, cont.

Pollution... 30, 55, 62, 69
Pollution Abstracts... 62
Proceedings in Print... 63
Rapra Abstracts... 64
Review of Agricultural Entomology... 65
Review of Plant Pathology... 66
Robotics Abstracts... 67
Rubber... 64
Science Citation Index... 68
Selected Water Resources Abstracts (SWRA)... 69
Soils and Fertilizers... 70
Soil Science... 70
SOLIDSTATE... 71
Solid State and Superconductivity Abstracts... 71
Veterinary Science... 48, 75
Water Resources... 69
Weed Abstracts... 72
Wildlife... 73
Wildlife Review... 73
Zentralblatt fur Mathematik... 74
Zoological Record... 75

## 9. Bibliography


Atkin, P. and P. Seed. "The acquisition of conference proceedings at the British Library." Interlending Review 7 (1979): 47-51.

Brahmi, Frances A. "Verifying the elusive proceedings: a review of available sources." Medical Reference Services Quarterly 5 (1986): 1-11.

Culnan, Mary J. "An analysis of the information usage patterns of academics and
practitioners in the computer field: a citation analysis of a national conference proceedings." Inf. Proc. Man. 14 (1978): 395-404.

Culnan, Mary J. "Information usage patterns in the computer field: a citation analysis." The information age in perspective: proceedings of the Asis Annual Meeting 15 (1978): 89-92.

Drubba, Helmut. "Literature on meetings and conferences. New bibliographic aids published between 1982-1984" . Translated title: Die Tagungs- und Konferenzliteratur. Neue bibliographische Hilfsmittel der Jahre 1982 bis 1984." Abi Technik 4 (1984): 289-293.

East, John W. "Citations to conference papers and the implications for cataloguing." Library Resources & Technical Services 29 (1985): 189-194.

Frankland, Garth. "Conferences as serials." Serials 1 (1988): 53-54.

Haigh, P. A. "Conferences and their proceedings." NLL Review 2 ( 1972): 7-10.

Huby, Danielle and Claude Hurisse. "How IFP processes data concerning meetings." Iatul Quarterly 4 (1988): 215-222.

LaRussa, Carol J. "Stalking the elusive conference proceedings." DLA Bulletin 7 (1987): 12-15.

McGlasson, Sheila. "The characteristics of conference proceedings: an examination of their
bibliographic control and a discussion of the problems that this serial type poses to the librarian." Uk Serials Group Newsletter 5 (1983): 5-10.

Mills, P. R. "Characteristics of published conference proceedings." J.Docum. 29 (1973): 36-50.

Mount, Ellis, ed. The Role of Conference Literature in Sci-Tech Libraries. Vol. 9. Science and Technology Libraries. Binghamton, NY: Haworth Press, 1989.

Short, P. J. "Bibliographic tools for tracing conference proceedings." IATUL Proc. 6 (1972): 50-53.

Snow, Bonnie. "Online puzzles: conference papers and proceedings." Database 11 (1988): 94-103.

Unsworth, Michael E. "Treating IEEE conference publications as serials." Library Resources & Technical Services 27 (1983): 221-224.

White, Philip M. and Jerry W. Breeze. "Verifying Conference Proceedings." Research Strategies Fall (1987): 191-196.

Whitley, Katherine M. "Comparison of meeting indexes." Reference Services Review 15 (1987): 85-90.

Zamora, Gloria J. and Martha C. Adamson, eds. Conference literature: its role in the distribution of information: Proceedings of the Workshop on Conference Literature in Science and Technology in Albuquerque, New Mexico, Learned Information, 1981.